SAE EDGE™ RESEARCH REPORT

Unsettled Issues Regarding Communication of Automated Vehicles with Other Road Users

Sven Beiker, PhD
Silicon Valley Mobility

EDGE DEVELOPMENT TEAM

Professor Klaus Bengler, PhD, *Technical University of Munich: Institute of Ergonomics*
Josh Domeyer, PhD, *Toyota Collaborative Safety Research Center*
Doug Frasher, *Vehicle Designer*
Nikhil Gowda, *Renault-Nissan-Mitsubishi Alliance Innovation Lab, Silicon Valley*
Jake Harper, Zoox
Brian Ka-Jun Mok, PhD, *BMW Group Technology Office, Silicon Valley*
Joey Penniman, *Nuro*
Bryan Reimer, PhD, *MIT Center for Transportation and Logistics*
Andy Schaudt, *Automated Vehicle Systems, Virginia Tech Transportation Institute*
Danny Stillion, *IDEO*
Brad Templeton, *Robocars.com*
Jingyi Zhang, *Zoox*
Yi Zhang, *Pony.ai*

SAE INTERNATIONAL®
Warrendale, Pennsylvania, USA

About the Publisher

SAE International® is a global association of more than 128,000 engineers and related technical experts in the aerospace, automotive and commercial-vehicle industries. Our core competencies are life-long learning and voluntary consensus standards development. Visit sae.org

SAE EDGE™ Research Report Disclaimer

SAE EDGE™ Research Reports focus on topics that are dynamic, in which knowledge is incomplete, and which have yet to be standardized. They represent the collective wisdom of a group of experts and serve as a practical guide to the reader in understanding unsettled subject matter. They are not meant to provide a recommended practice or protocol. The experts have assembled as a community of practitioners to contribute and collectivize their thoughts and points of view; these are not the positions of the institutions or businesses with which they are affiliated, nor is one contributor's perspective advanced over others. SAE EDGE™ Research Reports are the property of SAE International and SAE alone is responsible for their content.

About This Publication

SAE EDGE™ Research Reports provide state-of-the-art and state-of-industry examinations of the most significant topics in mobility engineering. SAE EDGE™ contributors are experts from research, academia, and industry who have come together to explore and define the most critical advancements, challenges, and future direction in areas such as vehicle automation, unmanned aircraft, cybersecurity, advanced propulsion, advanced manufacturing, Internet of Things, and connectivity.

Related Resources

SAE MOBILUS® Automated & Connected Knowledge Hub
https://saemobilus.sae.org/automated-connected/

SAE Team

Frank Menchaca, Chief Growth Officer
Michael Thompson, Director of Standards, Information and Research Publications
Monica Nogueira, Director of Content Acquisition and Development
Beth Ellen Dibeler, Product Manager
William Kucinski, Managing Technical Editor

Printed in USA

EPR2020023
ISSN 2640-3536
e-ISSN 2640-3544
ISBN 978-1-4686-0274-6

To purchase bulk quantities, please contact: SAE Customer Service

E-mail: CustomerService@sae.org
Phone: 877-606-7323 (*inside USA and Canada*)
+1-724-776-4970 (*outside USA*)
Fax: +1-724-776-0790

https://www.sae.org/publications/edge-research-reports

About the Editor

Sven Beiker, PhD, is the Founder and Managing Director of Silicon Valley Mobility, a mobility consulting and advisory firm that specializes in technical diligence, product roadmaps, and business models for mobility topics. The engagements span start-ups, investors, and corporations in the mobility and adjacent industries. In addition, Dr. Beiker is a lecturer at the Graduate School of Business at Stanford University, where he instructs students on strategies for start-ups and corporations in the field of automated, connected, electrified, and shared mobility.

With well over 20 years of experience gained during his tenure at McKinsey & Company, Stanford University, and the BMW Group, Dr. Beiker is dedicated to the future of the automobile and personal mobility. His mission is to improve sustainability, safety, efficiency, and convenience in vehicles and how consumers use them. He combines perspectives from technology, business, policy, and human factors.

Dr. Beiker also serves on advisory boards of several start-ups in the mobility space, as an advisor to the German American Chamber of Commerce in San Francisco, and as an advisor/co-editor to the Lecture Notes in Mobility of Springer Science+Business Media. He is a continuing contributor to the SAE EDGE Research Report series.

contents

Unsettled Issues Regarding Communication of Automated Vehicles with Other Road Users

Abstract

The impending deployment of automated vehicles (AVs) represents a major shift in the traditional approach to ground transportation; its effects will inevitably be felt by parties directly involved with vehicle manufacturing and use (e.g., automotive original equipment manufacturers (OEMs), public transportation systems, heavy goods transportation providers) and those that play roles in the mobility ecosystem (e.g., aftermarket and maintenance industries, infrastructure and planning organizations, automotive insurance providers, marketers, telecommunication companies).

The focus of this SAE EDGE Research Report is to address a topic overlooked by many who choose to view automated driving systems and AVs from a "10,000-foot perspective:" the topic of how AVs will communicate with other road users such as conventional (human-driven) vehicles, bicyclists, and pedestrians while in operation. This unsettled issue requires assessing the spectrum of existing modes of communication—both implicit and explicit, both biological and technological—employed by road users today.

NOTE: SAE EDGE Research Reports are intended to identify and illuminate key issues in emerging, but still unsettled, technologies of interest to the mobility industry. The goal of this report is to stimulate discussion and work in the hope of promoting and speeding resolution of identified issues. SAE EDGE™ Research Reports are not intended to resolve the challenges they identify or close any topic to further scrutiny.

SVEN BEIKER, PhD
Silicon Valley Mobility

Edge Development Team

Professor Klaus Bengler, PhD, *Technical University of Munich: Institute of Ergonomics*
Josh Domeyer, PhD, *Toyota Collaborative Safety Research Center*
Doug Frasher, *Vehicle Designer*
Nikhil Gowda, *Renault-Nissan-Mitsubishi Alliance Innovation Lab, Silicon Valley*
Jake Harper, *Zoox*
Brian Ka-Jun Mok, PhD, *BMW Group Technology Office, Silicon Valley*
Joey Penniman, *Nuro*
Bryan Reimer, PhD, *MIT Center for Transportation and Logistics*
Andy Schaudt, *Automated Vehicle Systems, Virginia Tech Transportation Institute*
Danny Stillion, *IDEO*
Brad Templeton, *Robocars.com*
Jingyi Zhang, *Zoox*
Yi Zhang, *Pony.ai*

ISSN 2640-3536

Introduction

The automation of driving is arguably one of the most intensely discussed topics in industry, academia, regulation, media, and the general public when it comes to the future of mobility and transportation. The overarching goal is to move people and goods more safely and efficiently and potentially at a lower cost. While many concepts have been shown and product announcements have been made, it is still unclear, when and (importantly) where those often called "driverless" vehicles might be deployed. However, while much of this challenge may lie in finding appropriate technology solutions, another important aspect is how such vehicles would—and ultimately will—interact with humans. Those include people inside and outside of the vehicle (i.e., passengers and any other human road users). This report identifies unsettled issues regarding such interactions of a "driverless" vehicle with other human road users. Throughout this report, the term "driverless" refers to the operating state of SAE Level 3 and 4, where a vehicle driver is not required to take control of the automated vehicle (AV) while it is operating in an automated mode [1].

To visualize pertaining aspects, the reader should imagine a scenario where a car approaches a crosswalk that a pedestrian is about to cross. Many people would suggest that "eye contact" is a very important method to communicate intent between those road users to ensure safety, albeit research has shown that it is not as effective as one might think [2]. And still, the question becomes how such intent can be communicated if a vehicle is driverless and eye contact or gestures—and other types of conventional nonverbal communication—are not possible, whether as an actual (or inadvertent) exchange of information or to provide road user reassurance (which also matters).

Much of those pertaining issues regarding external communication for AVs has already been discussed and many concepts have been shown and tested in public. Key aspects of communication challenges have already been identified, sometimes through research and unexpected situations. Those encounters and findings of various groups contribute to an "unsettled" (or "yet to be settled") issue in terms of how human-AV interactions should be designed. While there are groups (such as SAE International standards committees) currently investigating AV interaction solutions, as more AVs are tested, it becomes clear that even more focus regarding those concepts is urgently needed. This is not only to maximize safety and trust of road users toward AVs, but also to test such vehicles under real-world conditions with real-world interactions [3].

Some of the most general and also most debated questions are, for instance, what an AV should actually communicate to whom and in what way. A great deal of research, evaluation, and demonstration has already been conducted in this field, be it in industry, academia, or government, and initial standardization efforts have emerged. However, as appreciated as standardization efforts might be, several experts have started to voice concerns that it might be too early to define what such human-AV interactions should look (or sound) like. There can be a risk that, short of better knowledge, standards that are pushed too early might preclude opportunities to come up with even better solutions.

Many aspects of clarity, efficiency, practicality, and so forth need to be considered at this intersection of technology and psychology as more questions seem to arise the further such work progresses.

Therefore, in context of this SAE EDGE Research Report, a contemporary perspective seems to be in order. What questions are currently debated in this field and which of those require prioritization and standardization? This report seeks to provide clarity concerning the most pressing questions regarding external communication for AVs. This also means that this report does not aim to provide definitive answers. Its inherent purpose is to have practitioners agree on "asking the right questions" and providing directions that ultimately lead to answers.

External Communication for AVs: Motivation

As previously mentioned, a great deal of progress has occurred in the field of AVs. Concept demonstrations, test programs—even public pilots—all contribute to a very rich development environment. The community of forward-thinking researchers, engineers, social scientists, regulators, and others have learned from experience what it means when AVs encounter other road users without predefined rules or practices.

Some of those encounters have led to tragedy, such as a fatal accident in Tempe, AZ, in March 2019 when a pedestrian was fatally struck by an AV tested under human supervision. Whatever the causes and responsibilities were in this case, it is not known if the AV had external communication means beyond that of a conventional vehicle (i.e., it would not indicate whether it "sees" the pedestrian, whether it would let the pedestrian cross the street in front of it, or whether it would stop). Of course, one might argue that conventional vehicles also don't communicate with pedestrians (the drivers do, through the use of horns, turn signals, flashing lights, gestures, etc.), or that this particular AV itself was supposed to be operated or monitored by a human test driver. But still, this case—like many other accidents involving pedestrians—shows that additional external communication from the vehicle might have saved lives. And at the same time, one needs to ask the question if or how it might be possible to communicate information like "I will keep driving ahead" and/or "I don't perceive an object" to the right recipient. Those questions are the motivation for this report.

To expand on that motivation, one only needs to consider the rising numbers of pedestrian fatalities in the United States (US). From 2007 to 2016, they tragically increased from about 4,700 to almost 6,000 [4]. Of course, human error such as speeding, distraction, incapacitation, driving under the influence, and the like are by far the leading cause of traffic accidents. Current advanced driver-assistance systems have already proven to significantly reduce these events, and AVs have the potential to reduce them even further. Yet those numbers still show a scale and criticality of human-vehicle encounters in today's traffic, which needs to be addressed for future transportation systems as well. Compounding this issue—for instance—are the respective stipulations in public policy and guidance related to the interaction of pedestrians, bicyclists, and vehicle drivers:

1. **NHTSA Pedestrian Safety—10 Walking Safety Tips:** "7. Never assume a driver sees you. Make eye contact with drivers as they approach to make sure you are seen" [5].
2. **California Driver Handbook:** "If a pedestrian makes eye contact with you, they are ready to cross the street" and "Making eye contact with a driver does not mean the driver will yield the right-of-way" [6].

These two examples provide an indication of how the communication of vehicle drivers and other road users is organized and, to some extent, regulated today. When considering how communication should be organized in a future with driverless vehicles, one also needs to think about the public sentiment related to such a setting. For instance, one recent global consumer survey found that 48% of consumers feel a sense of fear toward "self-driving" cars, 46% feel anxiety, and only 32% feel trust [7]. Other studies also provide evidence on the importance of trust in the user's acceptance of an AV [8, 9]. All this seems to support the argument that pedestrians might very well be unsure, concerned, and uncomfortable when interacting with an AV. This might make matters even worse if such sentiment leads to even further misunderstandings and miscommunication between humans and AVs. In this context, some recent research shows that automation may shift the cause of accidents because it does not perform the driving activity in the same way road users are used to (and might expect from a "self-driving" car) [10].

In summary—with AVs currently entering public traffic as test vehicles and soon for public use, and the concern that miscommunication between road users and AVs could contribute to traffic accidents—the purpose of this report is to raise questions regarding external communication for AVs to ensure safe and efficient deployment of this promising technology.

Prior Work on External Communication for AVs

Table 1—while not exhaustive—summarizes key research and ideation contributions in the field of external communications for AVs. (For further reference, recent publications should be studied and considered [11].) Currently, external communication is a wide and rather unstructured field, ranging from academic research to corporate R&D to design concepts. All of them have their relevance. However, many reviewers of this report have reiterated that there is no common trend for how external communication should be designed for maximum safety, efficacy, and acceptance—hence the motivation of this report.

Table 1 reflects a sample of the current research into external communication for AVs and highlights the breadth, interest, and inconclusiveness in the field. To provide more context for this situation, Figure 1 shows a general model of the vehicle-to-other-road-user interaction, which highlights the complexity and breadth of such an exchange. In addition, the following figures show examples for intent communication of AVs, including symbols (Figures 2, 3), projections (Figure 4), text (Figure 5), and facial expressions (Figure 6). While those examples are all visual expressions of the vehicle's intent, there are also several examples for more types of expressions (e.g., auditory) that are discussed in various studies and concepts. However, visual means seem to be the most actively researched.

One can say that as much as there is a broad interest in the field, its is difficult to point to a clear direction or even recommendation on how to design—let alone standardize—the human-AV interaction for future AVs.

One can also easily turn to social media to find substantial interest from the general public on this topic. For instance, an informal, nonscientific video study from 2016 documents how bystanders react when encountering an AV [13]. The producers of the video stage a mild collision between a pedestrian and a (seemingly) automated road vehicle, apparently because no effective communication between the vehicle and pedestrian has been established. This video—with its 682,377 views (at time of publication)—clearly shows that society is interested and that the need for effective communication between AVs and other road users is required to ensure safety as this new mode of transportation is deployed in public.

In addition to work on the implementation side of AVs, there are also standards activities, which might eventually lead to the regulation of external communication for AVs. Table 2 provides an overview of several evolving standards (by SAE International and by ISO) and the Working Party on Lighting and Light-Signalling (GRE) Autonomous Vehicle Signalling Requirements (AVSR) taskforce at the United Nations Economic Commission for Europe (UNECE). It is important to emphasize their "evolving" nature as they are still relatively early initiatives by working groups aiming to standardize how the intent and state of an AV should be communicated to other road users. It should also be noted that the respective efforts thus far are not called "standards" and are instead given labels such as "recommended practice" (SAE J3134), "technical report" (ISO TR 23049, 23720, 23735), and "taskforce" (UNECE GRE). Those terms describe their intended purpose in such that they provide early guidance and reporting from those respective working groups that are ultimately aiming for standards, which are likely to be considered by regulators and may guide future

TABLE 1. **Overview of prior research and ideation projects related to external communication for AVs (not exhaustive).**

Title	Organization	Year of activity/summary
	RESEARCH STUDIES	
Automated Vehicle-Pedestrian Communication	University of Wisconsin, MIT AgeLab, Toyota (Figure 1)	Ongoing—Suggestions to design vehicle automation kinematics for communication in conjunction with external lighting signals
AV Intent—Automated Vehicle Communication and Intent with Shared Road Users	US DOT/NHTSA, University of Michigan and Westat	Ongoing—Study to determine how AVs communicate intent to other road users most effectively
Lighting signals on car driven by human in a "seat suit"	Ford and Virginia Tech Transportation Institute	Ongoing—Study to test method for self-driving vehicles to communicate their intent to pedestrians, human drivers, and bicyclists in an effort to create a standard visual language
interACT	Consortium of European corporations and academic institutions	2020—Designing cooperative interaction of AVs with other road users in mixed traffic environments
Vehicle-to-Pedestrian Communication Displays for Autonomous Vehicles	Duke University	2016—Comparison of various methods to effectively display vehicle-to-pedestrian street crossing information
Safe interaction between cyclists, pedestrians, and automated vehicles	Institute for Road Safety Research (SWOV), the Netherlands	2016—Overview of current knowledge about the interaction of pedestrians and cyclists with (partially) automated vehicles, identification of further research needs
CityMobil2	Consortium of European public/private partners	2016—Study on the implementation of AVs including survey on auditory and visual signals from AV to pedestrians
Autonomous vehicles' interaction with pedestrians (AVIP)	Chalmers University, Sweden	2015—Investigation of pedestrian-driver communication and development of a vehicle external interface
ÆVITA	MIT Media Lab	2012—Study on the design of biomimetic vehicle-to-pedestrian communication protocols for autonomously operating and parking on-road electric vehicles
	IDEATION CONCEPTS	
External communication panels	Drive.ai	2019—External communication panels for AVs to properly convey messages to other drivers, pedestrians, cyclists, and everyone else on the road
Autonomous vehicle notification system (patent # US010152892)	Lyft (Figure 5)	2018—Patent granted for a method used in AVs to determine the location and setting of other road users and to notify them (e.g., that it is safe to pass or that it is yielding)
e-Palette Concept	Toyota	2018—Concepts to indicate the use of an AV (i.e., for passenger or goods transportation)
Virtual eyes	Jaguar Land Rover	2018—Concept study to determine how much information future self-driving cars should share with users or pedestrians to ensure that people trust the technology
Vehicle-to-pedestrian communication (F015, Smart Vision EQ Fortwo)	Daimler	2015, 2017—Concepts for the communication of AVs primarily to pedestrians; allowing for safe street crossing and facilitating shared use
Smiling car	Semcon (Figure 6)	2016—Concept where self-driving cars interact by smiling to show that it will stop for pedestrians and allow for safe crossing
Pedestrian notifications (patent # US009196164)	Google (Figure 3)	2015—Patent granted for a device to notify a pedestrian of the intent of a self-driving vehicle; visual, auditory, physical signaling devices are covered
IDS Concept	Nissan	2015—Concepts with exterior lights and displays conveying to pedestrians and others the vehicle's awareness of its surroundings and signals its intentions

FIGURE 1. **Transactional model of communication adapted for vehicle-other-road-user interaction (University of Wisconsin-Madison/Toyota) [12].**

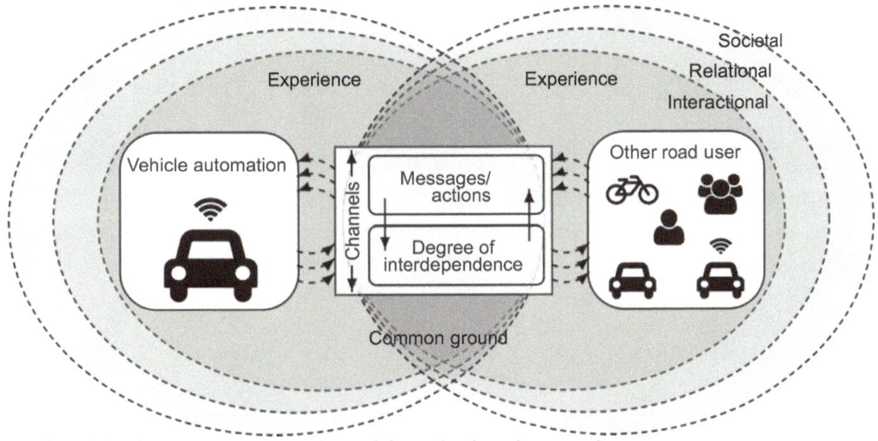

FIGURE 2. **Concept for an automated vehicle displaying advisory information as symbol.**

FIGURE 3. **Example for vehicle-to-pedestrian communication: an AV shows different symbols and signs to inform pedestrians when it is safe to cross the street (Google).**

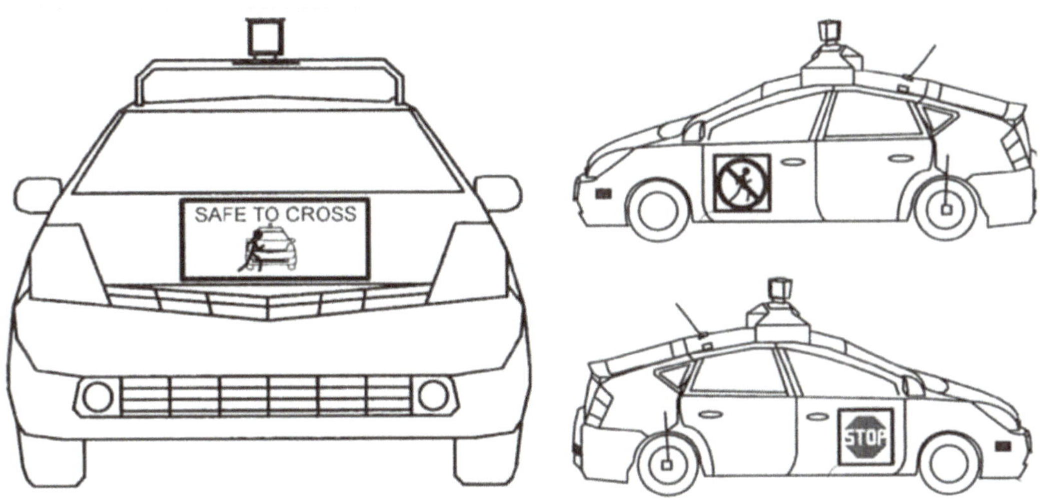

FIGURE 4. **Example for vehicle-to-pedestrian communication: an AV projects a virtual crosswalk as a sign that it lets the pedestrian cross the street in front of it.**

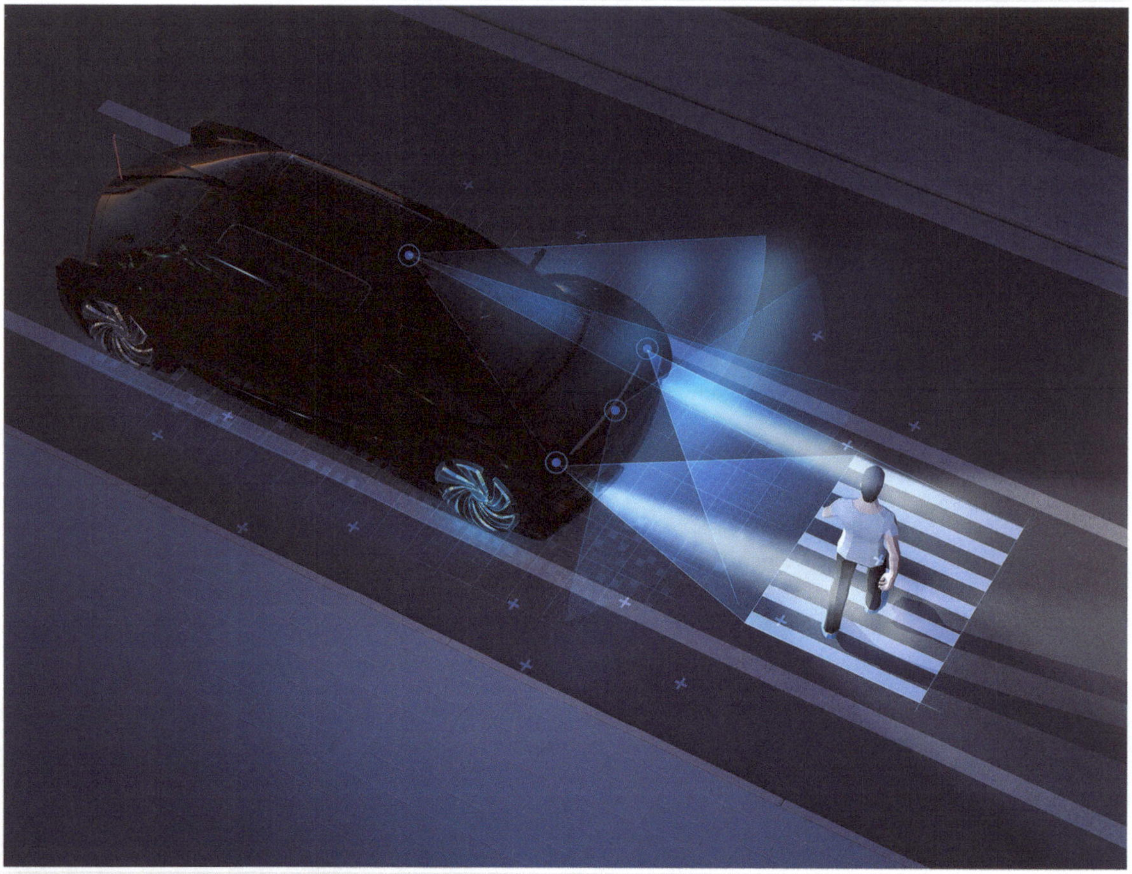

Chesky/Shutterstock.com

FIGURE 5. **Example for vehicle-to-pedestrian communication: an AV displays text message to pedestrians to indicate its intent to wait for them to cross the street in front of it (Lyft, 2018).**

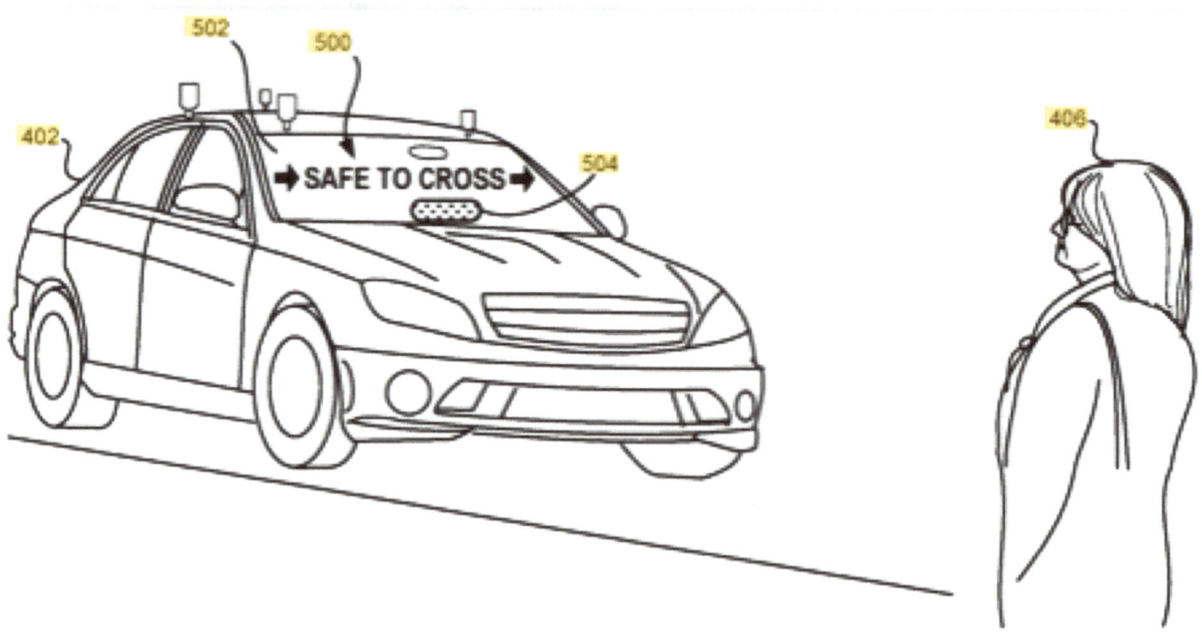

Reprinted from U.S. Patent 10,152,892 B2. United States Patent and Trademark Office (USPTO)

FIGURE 6. Example for vehicle-to-pedestrian communication: an AV shows a "smiling face" to pedestrians to indicate its intent to wait for them to cross the street in front of it (Semcon).

governmental rules such as the Federal Motor Vehicle Safety Standards (FMVSS) in the US or UNECE on an international level.

With those initiatives already underway, this report is also intended to provide further input to the respective working groups, in a sense as an additional "cross-check" and as an update on other parallel activities. By no means is this report intended to replace any of those initiatives, and it is important to note that insights from those working groups were included in the material provided herein.

An important aspect that unites those initiatives is that they all focus on visual communication. The SAE and ISO publications show that in their respective titles and the UNECE AVSR taskforce stated, as an outcome of the group's 5th session in March 2019, that "… the group concluded, that it should be a visible function […] this does not exclude in further discussions that audible signals […] may be taken into account" [14].

The initiatives are also united in their purpose that the intended standardization would be an important service to society. By standardizing the general implementation of such interaction concepts, road users are not left confused trying to interpret a variety of different communication efforts during critical encounters. For example, one can only imagine the problems arising from a pedestrian trying to cross the street while trying to interpret six different AV signals in addition to intersection signals, pedestrian signals, and the movement of road users/vehicles. This would be too much information to perceive and process, most likely leading to delayed reaction times or incorrect decisions. That is why the value of some level of standardization is a human factors safety issue and outcome.

Therefore, at a very top level of all those research, concept, and standards activities, two key aspects can already be identified as important for this report and beyond:

1. Visual means are seen as the most critical form of external communication for AVs (hence the further focus of this report).
2. External AV communication is a very interdisciplinary subject touching a broad spectrum of issues and solutions such as technology, psychology, social science, human factors, architecture, city planning, regulatory frameworks (including environmental regulation), insurance systems, and potentially other fields (e.g., business).

This second aspect makes it clear why the topic of external communication for AVs needs to be addressed on several levels and respective working groups, such as the ones driving the initiatives listed in Table 2. Those groups, if they aren't already, need to be comprised of experts from several different disciplines—well beyond engineering—which also implies a timeline to implementation that might be longer

TABLE 2. Overview of standardization activities related to external communication for AVs (non-exhaustive).

Title	Organization	Year of activity/summary
Ergonomic design guidance for external visual communication from automated vehicles to other road users (ISO/AWI TR 23735)	ISO	Ongoing—forthcoming technical report that encompasses design parameters of external signaling systems used by single mode L4/L5 Automated Driving System-Dedicated Vehicles (ADS-DVs), as defined by SAE J3016. Guidance is given for passenger cars (including sport utility vehicles and light trucks) and commercial vehicles (including heavy trucks and buses), as well as derivatives of them that carry or do not carry compartments for occupants (i.e., driver or passengers)
Road Vehicles—Methods for evaluating other road user behavior in the presence of automated vehicle external communication (ISO/AWI TR 23720)	ISO	Ongoing—more information forthcoming
Taskforce Autonomous Vehicle Signalling Requirements (AVSR)	UNECE	Ongoing—Task force to determine safety requirements for AVs to provide signals to indicate their status and to communicate their next intended actions
Automated Driving System (ADS) Marker Lamp (SAE J3134)	SAE International	2019—Recommended practice to provide guidance in the development of a standard for ADS lamps that is harmonized and recognized internationally to avoid confusion and complexity
Ergonomic aspects of external visual communication from automated vehicles to other road users (ISO TR 23049)	ISO	2018—Technical report to provide guidance for developers of visual external communication systems for AVs

than other, more technology-exclusive initiatives. For the same reason, the experts contributing to this report were also selected from different backgrounds such as engineering, design, human factors, public policy, and others.

Contributors to this report emphasized the need for implementation strategies in the industry (i.e., how compliance can be achieved) so that effective communication for maximum safety of all road users can be reached. And in parallel, education strategies need to be addressed (i.e., how penetration in society can be achieved) so that a common and correct understanding of external AV communication systems is realized. In this context, the term "learnability" was also used by experts as it becomes apparent that road users (i.e., all of us) will probably need to learn new interaction patterns in traffic with vehicles that are not driven by humans. In the future, conventional communication and signaling might not work anymore. However, as the following sections will discuss, experts diverged on whether interactions with an AV should be different from conventional patterns.

General Topics of Debate Regarding External Communication for AVs

Before going into further detail, general topics that emerged from the expert interviews regarding external communication for AVs need to be addressed. One key insight, which was surprising to the editorial team of this report, was the debate on whether such vehicles should have automation-specific external communication means at all. Some of the contributors stated they do not think an AV needs intent lighting. (The question was aimed at "AV intent lighting" because of the study's original scope, but general conclusions could be drawn from the pertaining discussion with the experts.) However, as binary as the question appears, there was much discussion surrounding these two options. While the question was specifically aimed at intent lighting, which is only a subcategory of external communication for AVs, the answers still carry weight for the broader discussion as it indicates a direction for how experts ascertain the general need for automation-specific communication means.

Several experts emphasized that AV intent lighting should not be different from conventional (human-driven) vehicles. For example, motor vehicles on public roads are required to have turn signals, brake lights, reverse lights, and—if applicable—warning lights and sound (e.g., emergency vehicles). Those examples illustrate that intent lighting (and in some cases sound) is already well established in public traffic. In that context, experts pointed out that it is most important to acknowledge that those communication means have become intuitive to road users through repeated encounter and are therefore easily understood. Those experts therefore emphasized that it is imperative that well-established external communication means are used by AVs for the same purpose (e.g., to indicate the intent to turn with signals or to slow down with brake lights). They further maintain that if additional communication means are considered, then they have to integrate (and not interfere) with existing schemes to avoid road-user confusion.

Overall, the opinions from the experts can be summarized in three categories:

1. Some kind of intent communication is required, but it is unclear what type and detail
2. No additional AV-specific intent signaling should supplement conventional (human-driven) vehicles intent signaling
3. Supplemental external AV communication means should be implemented—and in such a case, conventional (human-driven) vehicles should implement them as well

In particular, experts are very specific regarding the needs to integrate all types of AV intent communications with existing means of signaling. Related to this, experts also pointed out that "intent communication" needs to be discussed in much more detail apart from "state indication," which will be discussed later in this report.

General Perspectives on Whether AVs Should Have External Communication Means

Additional External Communication Exclusively Installed on AVs (Automation-Specific)

Some kind of intent communication is required, but it is unclear what type and detail.

Many of this report's contributors pointed to research, such as that summarized in "Prior Work on External Communication for AVs," that suggests a need for automation-specific communication to other road users in order to ensure safety and efficiency of public road traffic. Some experts also emphasized urgency in this as AVs are already being tested in public and are expected to launch for public use soon.

Despite some agreement that automation-specific communication is actually needed, the following question was also brought up by experts: "How this can be done most effectively covering all road users and all use cases?" This is of particular importance when considering the broad spectrum of road users from pedestrians and bicyclists to drivers and riders of motor vehicles, but also road users of different ages, sensory abilities, literacy, language skills, cultural backgrounds, and more.

Therefore, considering this spectrum of road users and variation of traffic situations, weather, and the like, experts also questioned at the outset of the discussion which communication channel would be most effective (e.g., visual, auditory, motion, or others). All contributors emphasized repeatedly the need for communication to be clear, unambiguous, and intuitive (to comprehend) to increase safety and not inadvertently lead to confusion and accidents.

Related, but potentially in conflict to this, is the suggestion that an implementation might need to be specific to the respective deployment scenario (e.g., a slow-moving vehicle in a dense urban area might communicate differently than a high-speed vehicle on a highway). With this, it becomes even more difficult to prescribe a universal solution in the short to medium term.

Overall, many experts agree that some form of communication is needed, but the following three limited examples show the complexity that leads to a lack of consensus regarding respective implement measures:

1. One AV and three human-driven vehicles approach a four-way stop at the same time. Humans typically wave their hand to decide who goes first (albeit this in itself is a debate among experts how well it works in practice), which might be difficult to recognize and actually learn for the AV; however, it can "creep" forward to communicate its intent to proceed (similar to what a human driver might do if hand-waiving does not solve the situation).
2. An additional challenge in an intersection situation might be that an AV likely detects "arrival time" at a different resolution compared to a human driver. For example, a human may view anywhere from 0 to 1 second to be the "same time," whereas an AV might use 0 to 0.1 seconds. Thus, the human may perceive the same time of arrival where the AV does not. This illustrates the importance of representing decisions in a human-like fashion in mixed traffic scenarios [15].
3. A pedestrian in a crosswalk might behave differently in different cultures or regions. Even in the same geographical region substantial differences in behavior might be found (e.g., traveling from the surrounding countryside to a downtown area). The intent signaling of an AV may be required to adapt to this behavioral phenomenon and research supports the need to consider cultural and regional differences [16].

While many agree that some kind of intent communication for AVs would be important, one also needs to ask the question whether those can be proprietary or individual solutions (i.e., different companies use different solutions and some companies none at all) or if those solutions need to be standardized to avoid confusion and ignorance among the intended recipients of such information. Therefore, standardization will play an important role in this issue. This will be a topic for another report.

Exclusive Use of Existing Vehicle Communication Means (Not Automation-Specific)

No additional AV-specific intent signaling should supplement conventional (human-driven) vehicles intent signaling.

FIGURE 7. Example for an automated mobility pod that is easily recognizable as an "atypical" vehicle (IDEO).

Reprinted with permission from IDEO

A percentage of contributors do not see the need for additional intent communication in AVs, which is related to the notion that existing communication means (e.g., turn signals and brake lights) should suffice. Furthermore, several contributing experts communicated that AVs should be self-evident and easily recognizable, which applies in particular to mobility pods like the one shown in Figure 7. Many experts suggest that if an AV is not self-evident, then at least its state (i.e., whether it is human-driven or driverless) should be communicated so that road users know that eye contact, for example, is not a functional method for negotiating a traffic situation. However, beyond this basic indication of AV state, experts caution to use any additional external communication.

A primary concern is that if AVs do not follow conventional behaviors or use additional signaling divergent from conventional vehicles in a mixed environment, there will likely be changes in the dynamics of traffic flow and other road users may become less vigilant, which can amplify risk and liability. Therefore, the goal must be that AVs eventually behave like human-driven vehicles and no additional means are needed to communicate their intent.

And an argument could be made that eventually all vehicles will be AVs and thus human-driven communication will eventually be the minority. The question then arises how to design for the medium and long term together? That also means, during a transition period, it might be beneficial to implement some indication to reassure other road users, which would help to build trust in the general public and lead to higher acceptance rates, more adoption and use of automation, and hopefully increase road safety through the reduction of human error-related incidents.

Additions to All, Including Human-Driven Vehicles

Supplemental external AV communication means should be implemented—and in such a case, conventional (human-driven) vehicles should implement them as well.

Several experts pointed to a mixed perspective regarding communication means for AVs; they suggested that signaling devices need to be considered holistically for all road vehicles. This was complemented by the findings from studies that human eye contact in traffic situations is not as effective as often thought because reflections, tinting, or misunderstandings make it difficult for road users to communicate through waving or nodding [17, 18]. In that sense, experts emphasize that if an electronic human-machine interface, or "eHMI," is beneficial for AVs, then it is most likely also beneficial for human-driven vehicles and should therefore be implemented in those as well to increase safety. However, any additional or changing interaction and/or communication of road users would need to be coupled with significant education as they would need a high degree of familiarity with changes prior to wide deployment to ensure road users have a clear and consistent mental model of the communication scheme.

Another question then becomes whether vehicles would need to be equipped retroactively with additional communication means if all vehicles (i.e., driverless and human-driven vehicles) use those. This consideration underlines the large scale that road traffic presents with close to 300 million registered vehicles in the US alone.

Purpose and Recipient of Communication

One of the most obvious questions regarding external communication might also be one of the most difficult ones to answer. It can be broken down as three related queries:

1. What needs to be communicated? (purpose)
2. Who should receive the communication? (recipient)
3. How should the information be communicated? (mode)

The following two sections will discuss the first two questions (purpose and recipient) in more detail. The third question (mode) will be pursued in greater detail as the core part of this report in "Modes of External Communication for AVs."

Intended Purpose Experts emphasize two particular aspects when it comes to external communication for AVs. The first one is safety, which should be the ultimate objective for all information exchange between road users. This leads to the need for clear, unambiguous communication that can be understood by all intended recipients in the same intended way. This will be discussed in more detail in "Clarity, Unambiguity, Efficacy." The second key aspect is to determine the intended purpose, which begins with the question if the "intent" of an AV should be communicated to other road users or if only its "state" should be communicated. To make this distinction clear, communication of an AV's "state" would indicate whether it is operating in an automated mode, whether it is operated by a human, or whether it is simply parked and not going to move at all. In comparison, communication of an AV's "intent" would indicate whether it is going to turn, slow down, depart, or perform other maneuvers in the near term. But longer-term intents might need to be communicated as well. One should only imagine an AV stopping at a very busy downtown crosswalk with an unabating stream of pedestrians. In such a situation the AV might need to indicate that it intends to move forward at some point and thereby negotiate the right of way with pedestrians. This again shows the complexity of situation that might appear clear in theory but can become rather intricate when put into real-world context.

While both types of information, state and intent, may be relevant to other road users' planning and actions, it may also be important to build public trust toward AVs. In that sense it might be necessary to provide other road users with information so that they feel safe and comfortable in the presence of an AV. Such information can be structured in three stages that can also be seen as a type of timetable (i.e., from general to particular information regarding a driving situation):

1. General behavior, performance, risks, etc. of a particular AV (common knowledge)
2. Current operation mode of an AV (state)
3. Near-term actions of an AV such as its immediate trajectory, which direction it will turn, etc. (intent)

The first stage—covered earlier—is mostly addressed by educating the public on AVs and how they operate. The second and third aspects (i.e. "state" and "intent") are, as pointed out at the beginning of this section, one of the fundamental questions regarding AV communication. Experts largely agree that the state should be communicated; however, communicating intent is a much more complex decision, especially in light of the discussion in "General Perspectives on Whether AVs Should Have External Communication Means."

Regardless of the ultimate decision of whether AVs communicate only their state or whether they also communicate their intent, it is important to note that experts often bring up the following purposes to be considered for AV communication. This list might not be complete and is only intended to provide communication purposes in a logical order:

1. Automated Driving System (ADS) on/off (i.e., vehicle driverless or driven by human)
2. Vehicle stopped/parked
3. Vehicle about to move
4. Vehicle about to stop
5. Vehicle about to turn (left or right)
6. Vehicle will stop for particular other road user (e.g., pedestrian in crosswalk)
7. Vehicle will yield for particular other road user (e.g., vehicle at intersection)
8. Vehicle about to creep forward in deadlock situation (e.g., four-way stop, busy crosswalk)
9. Passengers about to enter or exit (in particular for shared mobility purposes)
10. Vehicle in failure mode (also communication means failure)
11. Vehicle requires service/maintenance (in particular for shared mobility purposes)
12. Other situations to be identified once AVs become more prevalent
13. Warn others of impending hazard.

Intended Recipient The next key aspect of the communication from an AV is who the information is supposed to go to. One should not automatically assume that information should go to "all" other road users within range, as some content is rather specific. For instance, it might be of relevance and importance that an AV is about to move; however, the information that an AV yields to a particular person might be intended for a pedestrian but would be confusing to another vehicle in the broader vicinity. Therefore, it is very important to distinguish such cases and also to consider which individuals or groups should be addressed. Experts often mention the following individuals or groups as intended recipients of AV communication. Here again, this list might not be complete and is not in order of any particular priority:

1. Pedestrians
2. Bicyclists
3. Other motor vehicles (cars, trucks, motorcycles, etc.)
4. Ride-hailing users (in particular for shared mobility purposes)
5. Service personnel (in particular for shared mobility purposes)
6. Passengers within vehicle
7. Law enforcement, emergency vehicles
8. Other cases to be identified once AVs become more prevalent

One has to keep in mind that each of those categories is influenced by the respective scenario. The type of communication when tailgating an AV is different from the communication at a four-way intersection. Experts also caution that designing for (and even talking about) ideal situations is addressing only a small percentage of the problem statement they need to address. The challenge is communicating during edge cases that can be static (construction zone) or dynamic (yielding during rush hour). An interaction concept also needs to cover those rare and/or uncertain cases.

Related to this, it is crucial to determine if communication should be targeted toward individuals or groups. Here again, clarity and unambiguity are key as a vehicle might want to let all other road users know that it is operating without a driver but only wants to let other vehicle operators know that the vehicle is going to stop and passengers are about to exit. Therefore, it needs to be established which information is intended for all other road users without any priority and which information is only for a particular group or individual. For the latter, communication like "Waiting for you to cross" in Figure 5 is certainly very specific and it gets difficult to separate between intended and nonintended recipients.

This is also why experts maintain that a vehicle should only show its own intent and not be presumptive of what others might do. An easy way to think about this is how would a vehicle really know exactly what other road users are about to do and would it want to take responsibility for their actions? This is subtle but very important as it may be easy for road users to confuse text "okay to cross" with meaning that the vehicle is telling them the whole street is safe versus just the intent of the vehicle.

Communication Schematics

Related to the intended purpose and the recipient, it is also essential to determine how to express the intended purpose. The following sections discuss this in more detail.

Clarity, Unambiguity, Efficacy Several parts of this report have already stated that clarity and unambiguity are of prime importance for communication in traffic for all road users—not just AVs—even in a conventional, nonautomated setting. And still, with the introduction of AVs, with road users confronted with new situations and behaviors, with existing traffic rules and signs still applicable, it might be challenging to ensure that the intended recipient receives information, understands it correctly, and interprets it in the correct way for the intended purpose to take action. For AV communication, the following problems may occur, which might impact clarity and ultimately efficacy:

1. Communication not correctly sent
2. Communication path obstructed
3. Recipient distracted
4. Information not assigned to correct sender
5. Information not correctly understood
6. Information not correctly interpreted

In this context, the contributors emphasized the following objectives of primary importance to ensure the communication from an AV is clear, unambiguous, and effective:

1. Every intended recipient needs to understand information in the same and intended way
2. The amount of information needs to be limited, because road users cannot be expected to memorize or interpret a large amount of information or complex information
3. Communication needs to be aligned across different modes so that it becomes intuitive
4. External and internal communication need to be aligned (i.e., passengers might also need to be informed about external situation)

Feedback and Dialogue The previous examples and sections illustrate the complexity of what it might take to change the established communication scheme that road users navigate mostly intuitively today. Therefore, it is important to develop feedback mechanisms to ensure a set of information is received, understood, and interpreted in the intended way by the intended recipient. Such feedback mechanisms must be ubiquitous so that one can rely on the information coming through and that traffic dynamics allow for the natural behavior of previously established driving methods. Changing those dynamics bears the risk of confusion, disruption, and tragedy.

The concept of feedback mechanisms also leads to the question of how to communicate with an AV. This is not as unnatural as it might sound if one imagines a situation where an AV might signal that it will stop to let the pedestrian cross the street, but where the pedestrian might prefer to wait (and needs to signal that intent to the AV). This communication then becomes a dialogue, which is obviously very normal and occurs naturally between pedestrians and drivers of conventional vehicles. The question then becomes how such a dialogue can be implemented to ensure that information gets exchanged in the intended way between the intended participants.

Integration with Existing Communication Means It was mentioned earlier that conventional, human-driven vehicles use several different external communication means, which must be considered when proposing additional means for AV communication:

1. Headlights and taillights
2. Turn signals
3. Brake lights
4. Reverse lights
5. Hazard lights
6. Horns
7. Emergency lights and horns (e.g., police, ambulance, etc.)
8. School bus signs (e.g., stop sign, flashing lights, etc.)
9. Taxi signs (e.g., available/occupied, number of free seats, etc.)
10. Bus signs (e.g., route, service status, etc.)
11. Others

Those examples span a broad spectrum of communication means in which additional equipment need to be integrated; or even further, those examples offer opportunities, which additional communication schemes should leverage as much as possible. To illustrate this, the following intended purposes can be easily performed with the existing means from the previously discussed list (Table 3).

This illustrates that there are many use cases and communication means that overlap between conventional, human-driven vehicles and AVs. Thus, implementing new or additional communication means might not only be redundant, but also confusing. That is why many experts emphasize the use of existing communication means as much as possible—to the point at which additional means might not be necessary.

TABLE 3. **Examples for existing communication means to cover communication needs in AVs.**

Intended purpose	Use of existing communication means
Vehicle about to stop	Brake lights (maybe also forward facing)
Vehicle about to turn (left or right)	Turn signals
Not listed: pedestrian not obeying red light	Hazard and/or horns

Ultimately, the design of the communication scheme comes down to the question of which communication task needs to get solved and which communication mode should be chosen. While not all of those questions can be addressed in full detail here, more considerations regarding the communication mode will be discussed later.

General Key Questions Regarding External Communication for AVs

1. Do AVs need communication means in addition to conventional vehicles?
2. Should AVs communicate their intent, or only their state?
3. How can it be ensured that information from an AV is communicated correctly?
4. How can external communication for AVs be integrated with existing means?
5. How can coinciding messages be communicated with hierarchy (e.g., turning and stopping)?

General Recommendations

1. AV intent communication should leverage existing communication means (turn signal, brake lights, etc.) as much as possible.
2. If communication means are added to AVs, they need to be integrated with the means of conventional vehicles and potentially be added to those as well.
3. The scope of external AV communication should address the needs regarding intent (what is going to happen) and state (is it driverless or human-driven).

Modes of External Communication for AVs

The previous sections introduced different modes of communication for AVs, which generally align with human senses, as visible and audible AV communication is primarily intended to be received by humans. Therefore, vision, hearing, and touch/motion must be considered. By comparison, smell and taste are very unlikely to play a role in traffic. The following sections discuss different modes in more detail and also take into account combined modes. Visual communication will be addressed in significant detail according to experts' suggestions and in reflection of the prior work summarized in Table 1; however, details regarding the field of light will be covered in a later report.

Overview of Possible Modes

Visual and auditory communications are of prime importance in the conventional traffic setting. That means, road users primarily observe and recognize with their eyes and ears what happens around them and act accordingly. Other senses might also play a role but are only of secondary relevance.

Visual Visual communication in itself can have several different modes: simply a light that is on or off, a light flashing or blinking with a certain sequence, symbols or graphics, or displayed text. A balance between richness and comprehensibility of information needs to be found in between those visual modes. For instance, a rich full-text description of an AV's intent might be complete and even legally sound. However, it might be impossible for the intended recipient to comprehend such information over distance, in parallel to other sensory stimuli, or in the appropriate time. By comparison, a simple light might get to the intended recipient easily, but it might not be self-explanatory if a road user is not familiar with it. Table 4 provides examples and characteristics for different visual modes that compare such characteristics.

Some common aspects for all visual modes include directionality (i.e., a light can be shown only in a certain sector). For example, a forward-facing light is only intended for recipients in front of the vehicle. Applied to AVs, road users only need to be warned of a situation where the AV is approaching; road users behind the AV may not need to be informed that it is moving.

A common challenge for all visual modes listed is that visually impaired road users might not be able to perceive the information. While this is also already the case today with the conventional traffic setting of human-driven vehicles, one should ask the question how that can be improved when implementing a new type of transportation (i.e., introducing AVs). And even if someone has perfect vision, that person might not pay attention or may simply be looking the other way so that the communication would not reach its intended recipient.

Auditory Auditory is probably the second most used communication mode for road users, as it also plays an important role today. Here again, there are different modes within that category, such as a steady sound, intermittent sound, a sequence, or language. Table 5 gives an overview of those different auditory communication modes with examples and their respective characteristics. All examples are prevalent in traffic today and are often used to draw attention to visual information for road users who might be distracted or to provide alternative communication for the visually impaired, similar

TABLE 4. Examples and characteristics of different visual communication modes.

Visual communication mode	Conventional example	Characteristic
Steady light	Reverse light	+ clearly visible, even at distance + change of state is obvious – limited information richness
Intermittent, alternating, and/or cyclical light	Turn signal	+/– same characteristics as above + attracts attention, urgency
Symbol, graphic	Hand symbol "Don't Walk" on traffic light	+ enhanced information richness – risk for ambiguity – requires proximity to read
Text	"No turn on red"	+ maximum information richness – requires proximity to read – requires language/reading skill – requires processing time

TABLE 5. Examples and characteristics of different auditory communication modes.

Auditory communication mode	Conventional example	Characteristic
Steady sound	Horn for warning	+ attention attracted immediately – no specific recipient
Intermittent sound, single frequency	Reversing vehicle	+ urgency, affirmation can be expressed – no specific recipient
Sound sequence, varying frequency	Artificial electric vehicle (EV) sound	+ more pleasant, less annoying – ambiguity
Language	Crosswalk "Walk sign is on"	+ information richness – requires language skills – requires processing time

to the artificial sound implemented in electric vehicles when moving slowly or commercial vehicles when backing up [19].

The comparison of Tables 4 and 5 shows that auditory communication is an important complement to visual information to maximize the safety of all road users. Auditory means are often used to warn others if an incident is about to happen and that the information needs to reach the intended recipient in the shortest time possible. In that sense, it is a benefit that sound is not as directional as vision (i.e., it does not need to be on a particular side of a vehicle as long as it is loud enough) and it does not matter if the intended recipient is looking in the direction of the sender, if the view might be obstructed, or if the recipient might not be able to see at all.

While omnidirectionality and prevalence of sound are benefits to reach a recipient under basically any circumstance, experts also expressed concerns regarding this effect's "double-edged" nature. The problem of prevalent auditory communication in traffic would be noise pollution and an overload of this communication channel. In other words, this would lead to urban traffic (in particular) getting even louder—one might not be able to tell which sender is intending certain information for which recipient. To picture this, one should imagine a busy downtown intersection with several vehicles, bicyclists, pedestrians, and several AVs emitting a sound signal expressing "I will stop for you at the crosswalk." This might lead to a very unpleasant, confusing, and unsafe situation. Also, if the intended recipient of the auditory communication is inside another vehicle, it must be very loud to reach them (car makers are continuously working to soundproof vehicles to make the interior environment quieter).

Finally, one clear benefit of sound is that it can be tuned to be more assertive if the intended reaction of the recipient is not immediately accomplished. Sound can be tuned to be louder, harsher, more frequent, and so forth. All those are examples that are used by humans when honking, certainly at times inappropriately, but this can be imagined to be effective in some cases. And it should also be mentioned that regarding the directionality of sound, experts point to technology that can focus auditory information to specific locations (e.g., a specific pedestrian zone) and limit environmental impact of noise pollution at a wider scale. However, this area would need additional exploration.

Motion Motion is a communication mode that some experts suggest for AVs, which can be similar to gestures of a person. In that sense, if a vehicle tilts forward, backward, left, or right, one might assume that the vehicle is braking, accelerating, turning left, or turning right. In technical terms, one would say that the forces of inertia lead to pitch or roll motion of the body and an observer might anticipate a change in the vehicle's motion. With an active suspension, one can even proactively induce such pitch or roll motion in vehicle and therefore communicate its intent.

While such communication is ultimately still based on a visual concept, it deserves special consideration as experts emphasize the fact that whatever intent communication is chosen for an AV, it has to be aligned with the motion of the vehicle. That means, for instance, if an AV communicates that it is about to stop, then it is expected that the body tilts forward and not backward. This actually can be extended to the sound of the vehicle as it would be confusing if an AV communicates its intent to stop but the driving noise increases at the same time.

However, some experts maintain that technical knowledge is necessary to associate vehicle body pitch or roll with the correct change in motion. Yet, other experts point out that even if the average road user might not be familiar with vehicle dynamics, many have subconsciously learned certain vehicle motion behavior.

Other Possible Modes Beyond the most obvious communication modes (i.e., visual and auditory) and the possible addition of motion, one can also imagine even further solutions. Those might draw on connectivity, such as vehicle-to-pedestrian (V2P) communication. Figure 8 shows an example for a concept that would alert pedestrians of approaching vehicles that present a hazard.

This is just one example of additional modes for intent communication; others can be imagined with the use of communication technology. For instance, vehicle-to-vehicle (V2V) communication may also be included in some way (e.g., reassuring vehicle occupants that vehicles are indeed communicating through "connected status" indicator), which may help build occupant trust and confidence. This is yet another example of the broad range of interaction concepts that require a holistic approach outside and inside the vehicle.

While wireless communication solutions will not be discussed in more detail here, one should always keep in mind that clarity, unambiguity, and efficacy are most important to ensure safety in traffic. Therefore, less common and more complex solutions for intent communication should only be considered in special settings, which will probably require special education for the road users involved.

SAE International has been addressing respective solutions in several activities such as the J2945 series that covers dedicated short range communication (or "DSRC") systems, in particular the J2945/9 recommended practice "Vulnerable Road User Safety Message Minimum Performance Requirements" [20] and the forthcoming J2945/D recommended practice "Road user-to-Road User Courteous Communication" [21].

Comparison of Different Modes in External Communication for AVs The different solutions discussed in the previous sections should be compared with one another regarding their relative strengths and challenges (Table 6).

While these are rather one-dimensional insights, such as if a certain requirement can be addressed by one particular mode or not, it is also valuable to consider several requirements in conjunction. After all, the entire spectrum of characteristics needs to be covered. Therefore, it is of particular interest what the experts contributing to this report point to in their comparison.

Some experts describe the comparison of visual and auditory communication modes as a balance of urgency, range, bandwidth, and complexity that needs to be optimized based on a particular use case. Among the different visual

FIGURE 8. **Example for vehicle-to-pedestrian communication that warns the pedestrian of a vehicle presenting potential hazard.**

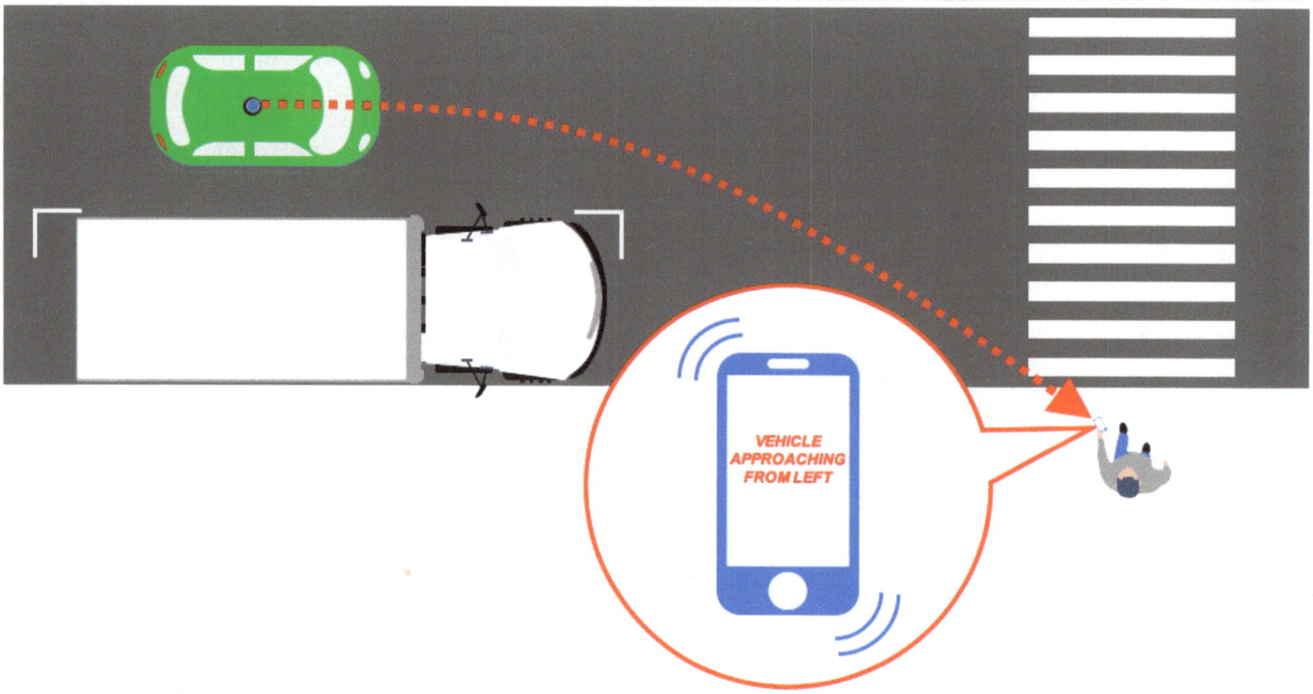

TABLE 6. **Comparison of characteristics for visual and auditory communication modes (relative comparison, non-exhaustive).**

Requirement	Preferred mode (V = visual, A = auditory)
Clarity	V: steady light (if recipient understands meaning)
Low risk of misunderstanding	A: steady sound (if recipient understands meaning)
Information richness	V: text
High content of information	A: language
Assertiveness	V: intermittent light, e.g., increased frequency
Forcing recipient to pay attention	A: steady/intermittent sound, e.g., increased volume
Penetration	A: any mode
Resistant to impedance, obstruction	
Range	V: steady or intermittent light
Information travels long distance	A: steady or intermittent sound
Directionality	V: any mode
Focus on individual recipient, group	
Efficiency	V: any mode
Limited channel load, no "pollution"	Motion: any mode
Intuitiveness	V: steady, intermittent light as conventional
Understood by prior experience	A: steady sound or sequence as conventional

modes for instance, text does not travel far but is information rich and can convey complexity. A steady light is the opposite, as it travels far but does not provide complexity. The question might be what optimal solution could be found in between those two points if complex information needs to be conveyed over a certain distance while ensuring clarity.

For auditory communication, the experts appreciate that it has a very high chance to reach the intended recipient. However, it might lack clarity and may reach many other (unintended) recipients, thereby leading to noise pollution, especially in an urban environment. Those aspects might limit audio to warnings and alerts rather than some sort of state or intent communication.

One expert compares the two primary communication modes as "auditory is push communication that gets through no matter what, and visual is pull communication that a road user needs to be actively looking for."

Summary

The most basic but still unanswered question regarding external communication for AVs repeatedly pointed out by the experts contributing to this report is

Do automated vehicles need any external communication systems in addition to those currently on conventional vehicles?

This report offers a broad spectrum of information, opinions, and suggestions provided by experts in the field. And still, this most basic question cannot ultimately be answered. The body of experts tend to agree that visual communication that expresses the state of an AV (operated driverless or by a human) bears some merit. They recommend providing essential information to other road users who are typically inexperienced in such human-AV interaction as AVs are launched into public traffic. They also emphasize repeatedly that clarity and unambiguity are of primary importance to avoid confusion and ensure safety.

SAE EDGE™ Research Reports

SAE EDGE™ Research Reports, like the present report on "Unsettled Issues Regarding Communication of Automated Vehicles with Other Road Users" are intended to push further out into still unsettled areas of technology of interest to the mobility industry. SAE launches these reports before attempting to form a joint working group, let alone a cooperative research program or a standards committee.

These reports are intended to be quick, concise overviews of major unsettled areas where vital new technologies are emerging. An unsettled area is characterized more by confusion and controversy than established order. Early practitioners must confront an absence of agreement. Their challenge is often not to seize the high ground but to find common ground. These scouting reports from the frontiers of investigation are intended merely to begin the process of sorting through critical issues, contributing to a better understanding of key problems, and providing helpful suggestions about possible next steps and avenues of investigation.

SAE EDGE Research Reports, therefore, are fundamentally distinct from the more formal working groups approach and far removed from the more mature research program and standard's development process.

Next Steps for External Communication for Automated Road Vehicles

This publication should be considered only as a first step toward clarifying the issues around external communication for automated road vehicles. The intention behind this and other SAE EDGE Research Reports is to start a dialogue among interested parties on important industry-wide topics that require further attention. The expectation is that these explorations of unsettled areas of technology will lead to the formation of working groups and, ultimately, committees that can address and resolve the issues they raise, producing a framework for developing a common vocabulary of definitions, best practices, protocols, and standards needed to support continued progress toward safer and more innovative products.

The expert discussions that gave rise to this publication demonstrated a great willingness on the part of the industry to define the terminology, procedures, and eventually enable the potential deployment and standards development for external AV communication. SAE International has demonstrated its lead in ADS and AV standardization, namely the rollout of the ubiquitous J3016 standard, "Taxonomy and Definitions for Terms Related to On-Road Motor Vehicle Automated Driving Systems," the J2735 "V2X Communications Message Set Dictionary," and J3018 "Safety-Relevant Guidance for On-Road Testing of SAE Level 3, 4, and 5 Prototype Automated Driving System (ADS)-Operated Vehicles."

This report on external communication for automated road vehicles identifies the following key topics for further pursuit, both through continued informal discussions among industry practitioners and through more formal working groups:

1. What is the intended purpose/information of the communication?
2. Who is the intended recipient of the communication?

3. How is the intended recipient supposed to react to the information?
4. How can this communication be implemented most effectively?

As the expert discussions and the review of prior work in the field almost unanimously suggest that visual modes play the most important role in traffic interactions, those aspects will be addressed in significant detail in a specific publication. Lighting installations are well established and apparently well suited as a communication mode for road users so that those deserve particular consideration. This topic will be covered in a future report.

Recommendations

Experts point out that one needs to deliberate thoroughly if there are actual technical reasons to implement automation-specific communication or if it is primarily to provide reassurance to road users and offer a substitute for human eye contact (which is not possible with an AV). If there is no technical reason, then educational matters regarding the operation, benefits, and risks of AVs might be more effective to ensure safety than implementing communication measures on a vehicle, which might be overlooked or misinterpreted.

There are still many unknowns regarding what an AV actually will be like (to date there are no public deployments except tests and early pilot programs) and what information other road users will need in order to share the road with them. Ultimately, all experts that contributed to this report point out that more research and pilot programs are necessary to achieve a better understanding and inform standards regarding external communication for AVs. While it seems plausible in theory that AV intent and state communication can greatly improve pedestrian comfort and reaction, experts caution not to rush to solutions until use cases and the level of detail to which other road users need specific signaling are understood. Then standards can be developed based on sufficient user study and field testing to implement a unified language across involved parties and products.

Because many concepts have already been shown with a variety of different communication types and modes, experts highly recommend aligning on facts regarding the efficacy of human-AV interaction with safety as the primary objective before moving toward standardization of measures (should research show that additional measures are necessary).

Definitions

ADS - Automated Driving System

AV - Automated Vehicle

AVSR - [UNECE GRE] Autonomous Vehicle Signalling Requirements

DOT - [US] Department of Transportation

EV - Electric Vehicle

FMVSS - Federal Motor Vehicle Safety Standards

GRE - [UNECE] Working Party on Lighting and Light-Signalling

ISO - International Organization for Standardization

MIT - Massachusetts Institute of Technology

NHTSA - National Highway Traffic Safety Administration

OEM - Original Equipment Manufacturer

R&D - Research and Development

UNECE - United Nations Economic Commission for Europe

US - United States

V2P - Vehicle-to-Pedestrian

V2V - Vehicle-to-Vehicle

Acknowledgments

Recognition should go first to all the participants, many of whom also provided feedback on the draft version of this publication. Without their input and initiative, this report would not have been possible.

Professor Klaus Bengler, PhD	Technical University of Munich: Institute of Ergonomics
Josh Domeyer, PhD	Toyota Collaborative Safety Research Center
Doug Frasher	Vehicle Designer
Nikhil Gowda	Renault-Nissan-Mitsubishi Alliance Innovation Lab, Silicon Valley
Jake Harper	Zoox
Brian Ka-Jun Mok, PhD	BMW Group Technology Office, Silicon Valley
Joey Penniman	Nuro
Bryan Reimer, PhD	MIT Center for Transportation and Logistics
Andy Schaudt	Automated Vehicle Systems, Virginia Tech Transportation Institute
Danny Stillion	IDEO
Brad Templeton	Robocars.com
Jingyi Zhang	Zoox
Yi Zhang	Pony.ai

The authors of this document together with the SAE Team responsible for its creation join in expressing their deepest appreciation to all of the individuals who contributed.

Sven Beiker, PhD
Silicon Valley Mobility

References

1. SAE International, "J3016: Taxonomy and Definitions for Terms Related to Driving Automation Systems for On-Road Motor Vehicles," https://www.sae.org/standards/content/j3016_201806/, accessed Nov. 11, 2020.
2. AlAdawy, D., Glazer, M., Terwilliger, J., Schmidt, H. et al., "Eye Contact between Pedestrians and Drivers," in *Proceedings of the 10th International Driving Symposium on Human Factors in Driver Assessment, Training, and Vehicle Design*, Santa Fe, NM, 2019, 301-307.
3. Rothenbücher, D., Li, J., Sirkin, D., Mok, B. et al., "Ghost Driver: A Field Study Investigating the Interaction between Pedestrians and Driverless Vehicles," in *2016 25th IEEE International Symposium on Robot and Human Interactive Communication (RO-MAN)*, New York, 2016, 795-802.
4. NHTSA, "Traffic Safety Facts: Pedestrians," https://crashstats.nhtsa.dot.gov/Api/Public/ViewPublication/812493, accessed Nov. 11, 2020.
5. NHTSA, "Pedestrian Safety," https://www.nhtsa.gov/road-safety/pedestrian-safety, accessed Nov. 11, 2020.
6. California DMV, "Driver Handbook," https://www.dmv.ca.gov/portal/driver-handbooks/, accessed Nov. 11, 2020.
7. Capgemini Research Institute, "The Autonomous Car: A Consumer Perspective," Capgemini, Apr. 2019.
8. Choi, J.K., and Ji, Y.G., "Investigating the Importance of Trust on Adopting an Autonomous Vehicle," *International Journal of Human-Computer Interaction* 31(10):692-702, 2015.
9. Lee, J.D., and Kolodge, K., "Exploring Trust in Self-Driving Vehicles through Text Analysis," *Human Factors* 62(2):260-277, 2020.
10. Bainbridge, L., "Ironies of Automation," *Automatica* 19(6):775-779, 1983.
11. Dey, D., Habibovic, A., Löcken, A., Wintersberger, P. et al., "Taming the eHMI Jungle: A Classification Taxonomy to Guide, Compare, and Assess the Design Principles of Automated Vehicles' External Human-Machine Interfaces," *Transportation Research Interdisciplinary Perspectives* 7:100174, 2020.
12. Domeyer, J.E. et al., "Vehicle Automation—Other Road User Communication and Coordination: Theory and Mechanisms," *IEEE Access* 8:19860-19872, 2020.
13. Self-Driving Vehicle HITS Pedestrian!!, Rich Ferguson, https://www.youtube.com/watch?v=7FpRNn_dcTE, accessed Nov. 11, 2020.
14. UNECE, "AVSR—5th session," Mar. 13, 2019, https://wiki.unece.org/display/trans/AVSR+-+5th+session%2C+2019-03-13, accessed Nov. 11, 2020.
15. Toyoda, H. and Domeyer, J., "Autonomous Vehicle Operation at Multi-Stop Intersections," U.S. Patent Application 20170057514A1, Nov. 21, 2000.
16. Li, J., Currano, R., Sirkin, D., Goedicke, D. et al., "On-Road and Online Studies to Investigate Beliefs and Behaviors of Netherlands, US and Mexico Pedestrians Encountering Hidden-Driver Vehicles," in *Proceedings of the 2020 ACM/IEEE International Conference on Human-Robot Interaction (HRI '20)*, 2020, Association for Computing Machinery, New York, 141-149.
17. AlAdawy, D. et al., "Eye Contact between Pedestrians and Drivers," in *10th International Driving Symposium on Human Factors in Driver Assessment, Training and Vehicle Design*, Santa Fe, NM, 2019.
18. Dey, D. and Terken, J., Pedestrian Interaction with Vehicles," in *Proceedings of the 9th International Conference on Automotive User Interfaces and Interactive Vehicular Applications*, Oldenburg, Germany, 2017, ACM, 109-113.
19. National Highway Traffic Safety Administration, "Minimum Sound Requirements for Hybrid and Electric Vehicles," Report No. DOT HS 812 347, Nov. 2016.
20. SAE J2945/9 Vulnerable Road User Safety Message Minimum Performance Requirements, 2017.
21. SAE J2945/D Road User-to-Road User Courteous Communication, (forthcoming).

Contact Information

EDGEresearch@sae.org.

www.ingramcontent.com/pod-product-compliance
Lightning Source LLC
LaVergne TN
LVHW070941210926
848482LV00003B/273